first of all
blowing the scarecrow down
autumn storm

Kyoraku

The Scare Crow

A Collection of Haiku & Senryu

Compiled & Edited by
Leroy Kanterman

The Scare Crow

ISBN 978-1-893959-00-2

Red Moon Press
P. O. Box 2461
Winchester VA
22604-1661 USA
www.redmoonpress.com

second printing

Translation of "first of all" by Harold G. Henderson.

Some of these poems appeared previously in *Frogpond, Geppo Haiku Journal, Haiku West, Modern Haiku, New World Haiku, Spin, The Essence of This, Woodnotes,* and The World Haiku Contest.

To the memory of a dear friend
and co-founder of The Haiku Society of America

HAROLD GOULD HENDERSON
1880 – 1974

Preface

The scarecrow, an ephemeral figure, serves as a poetic device to convey the gamut of human emotions, ranging in scarecrow haiku, for example, from the comic to the tragic.

Most haiku result from an inner response to outer experience. In scarecrow haiku, which for the most part are imaginary, the process is reversed. The poet is able to express his or her feelings by using this human surrogate, the scarecrow, to cover the range of humor and pathos.

In this collection from contemporary poets and early Japanesse masters, most of that range is represented.

— *Leroy Kanterman*

The Scarecrow in our Haiku

It seems that the image of a scarecrow is well placed for use in a form of poetry which seeks to link human nature with greater nature. Shaped like a person and intended to interact with the birds, a scarecrow spends day and night out-of-doors, has a kind of life which is attuned to the seasons and in which weathering provides a near-equivalent to aging. The very ease of its application to haiku, however, has made me wary of it. Growing up in farm country and living much of my adult life in rural areas, I have never seen a scarecrow used in a field. I am told that the scarecrow is making a comeback in New England but have not noticed this just over the border, in Eastern New York. My associations from experience are of scarecrows in gardens or ready-made in stores, restaurants or other commercial settings. The primary function of these scarecrows has seemed to be decoration, whimsy or the expression of some sort of domestic fantasy. Personally, this is just the sort of thing I would be pleased to keep out of my haiku. And yet, I don't sense that this is what others make of a scarecrow at all.

Perhaps it is the case with scarecrows that we have a kind of collection memory of a time when the world was not so overpopulated, when the survival of human beings, both the individual and the village, was always more at issue from day to day, moment to moment, than we perceive it to be today. It seems likely thatd the first scarecrows were quite the opposite of whimsical. They were probably meant, like most simple tools, to extend the hands of laborers who were in short supply. Some sense of our ancestors' need for the harvest which meant the difference between subsistence and starvation may still attach itself to the scarecrow image. Certainly many of us seem to see in the scarecrow's ragged state either our own frailties or the plight of homeless persons. On the whole, though, it seems rather ironic to invoke the image of extra hands in a world with billions of people.

Another aspect of the scarecrow which may inform its use in our poems is the spiritual, which mwe might tend to describe in our times as the superstitious. During most of the centuries in which scarecrows were in common use there was a greater faith in the imminent presence of spirits, in the home, the woods and fields. The birds were viewed as spirits, as were the seeds, rain, wind and the earth itself. Indeed, everything was viewed as having a kind of life in it, even a made-up thing like a scarecrow. Many a farmer must have looked upon the scarecrow in his field as a being, a comrade. In this sense, the scarecrow was a relative of those effigies of bygone times, such as those seen on the bows of sailing vessels or at the gates of an ancient Assyrian palace, whose role was one of spiritual guardian and ally. The use of scarecrows has declined since the late nineteenth century, partly for practical reasons relating to new agricultural methods, but its decline may also correspond in some degree to what Matthew Arnold described as the long ebbing of the "sea of faith" during the same period.

It seems probable that our haiku may also reflect some sense of the folk art aspects of the scarecrow, which is very consistent with the "nothing special" æsthetics of haiku. Because of our limited exposure to actual scarecrows, however, what we picture inour imaginations may have considerably less variety than the real things. We are likely to think of something along the lines of C. Frank Baum's scarecrow in *The Wonderful Wizard of Oz,* or even more likely, of the Ray Bolger characterization from the famous movie. This particular example is an Americanization of the European type of scarecrow, brought here by immigrant farmers. But there are many variations within that tradition, as well as the independently developed varieties native to, for instance, Asian and Native American cultures. Avon Neal and Ann Parker, in *Ephemeral Folk Figures*, provide dozens of striking photographs of unique scarecrow figures (a few of which are reproduced in these pages) and point out that many a scarecrow bears, intentionally or otherwise, a strong resemblance to its creator. In the making of scarecrows, many people who otherwise would have considered themselves hard-bitten realists produced works of art, perhaps for the only time in their lives.

Occasionally, in English-language haiku journals, one comes across criticism of haiku which anthropomorphizes some aspect of non-human nature. In general, I am in sympathy with this concern but the very existence of the scarecrow image suggests that there can be no immutable rule. The scarecrow is, after all, an amalgam of natural and manufactured objects and represents an obvious interface between nature and man's attempts to influence and channel nature. Is the scarecrow primarily reflective of human nature because it is made by a human being for human purposes or is it part of nature because of what the wind, rain and sun make of it in the courses of the sesons? Perhaps the question itself is an artifact of our culture. The distinction between nature and human nature may not be as sharp as we perceive it to be.

I suspect taht most of our poems about scarecrows will be desk haiku, based on either remembered things or on projections. The scarecrow in our haiku is a parallel construction to the scarecrow in ancient fields, made in the image of ourselves, to extend our reach: a companion among spirits: an expression of the urge of practical people to make art. It is a potent tool for poetry, with strong suggestions of loneliness, poverty and patient endurance, of humor, humility and of the uncanniness of a human figure transfixed in the lengthening shadows or poised in the darkness.

—John Stevenson
Nassau NY
October 1998

The Scare Crow

looking for a place
to quietly dispose
the old scarecrow

Michael Cecilione

cold moon
    inside the scarecrow
    crickets singing

Michael Cecilione

walking with the wind
something left in the pocket
of the old scarecrow

Ross Figgins

new scarecrow
  I peer over the corn
    on tiptoe now

Ross Figgins

Also waving it's arm
the scarecrow's
shadow

Garry Gay

The scarecrow
in tall weeds
all a bloom

Garry Gay

in a snowy field
one lone scarecrow is guarding
a tree full of crows

Harold G. Henderson

and still he stands there
the scarecrow — in a field
not planted this year

Harold G. Henderson

autumn breeze
the scarecrow's shadow
   grows thinner

eric l. houck, jr.

night wind
  scarecrow
shifting on his post

eric l. houck, jr.

winter blast
and the scarecrow loses
  another straw

Ken Hurm

my neighbor
admiring my scarecrow
steps on the corn shoots

Ken Hurm

staring
    at the wild geese
scarecrow

Leroy Kanterman

on the scarecrow
the autumn wind has settled
  a coat of dust

Leroy Kanterman

moonlit cornfield
how softly an owl
skirts a scarecrow

Emily Romano

rising toward stars
ignited fragments
of the scarecrow

Emily Romano

country farm market
a scarecrosw waving from
  the pumpkin field

Bruce Ross

sunny august day
the tee-shirted scarecrow enveloped
by tomato plants

Bruce Ross

a cricket
lending the scarecrow
a voice

John Stevenson

ginko —
our hostess leads us
toward a scarecrow

John Stevenson

October wind
scarecrow bends
   toward the crows

Tom Williams

scarecrow—
the wind turns him
to face the east

Tom Williams

From the Japanese

translations by
Hiroaki Sato

The Three Types of *Kagashi*

1) A representation of an agricultural deity made to pray for a bumper crop or to prevent insects and animals from damaging the crop; it may be a deity-shaped doll, a simple paper-tablet, or a shinto rope called shime. Some may be made with a plow, a sickle, and a sedge-hat. Today, *kagashi* normally refers top any of the representations in the last category, especially a man-sized doll made of bamboo, straws, or discarded clothes, wearing some kind of hat, with an arrow or bow held in one or both hands, and standing on a single stick. The names vary by region, in some caled *odoshi* ("threat") or *odorokashi* ("alarm").

In the *Kojiki* (*Record of Ancient Matters*, complied 712AD), what is today known as *kagashi* is described thus: ". . . he is still called Sohodo in mountain paddies today. This deity can't walk but knows everything under heaven."

2) Foul-smelling things, such as rags, hair, meat, and fish heads, which are burnt and strung together. The now almost-universal name of *kagashi* actually means "let them smell this" and derives from this category of devices.

3) A sound-making objectd, such as a clapper, or a dead bird or animal hung to scare birds and animals.

In *renga* and *haikai*, any of these objects, called variously *kagashi*, *tori-odoshi* (scare-bird), *odoro-kashi*, ("scare-deer") and hajiki ("repellent") has been treated as a *kigo* or *kidai* for autumn.

— *Yamamoto Kenkichi*

midnight frost:
to sleep I borrow
sleeves of the scarecrow

Bashō

a sound . . .
alone a scarecrow
has fallen down

Bonchō

the autumn wind
moving a scarecrow
moves on

Buson

without his hat
no dignity's left
for the scarecrow

Buson

the water drained
the scarecrow's skinny shank
rises high

Buson

a cricket chirps
  in a sleeve
of the scarecrow

Chigetsu-Ni

the autumn wind
  penetrates
the scarecrow's bones

Choi

before royalty
the scarecrow doesn't remove
his plaited hat

Dansui

his hat fallen
  the rain
beats on the scarecrow

Hagi-Jo

even the scarecrow
turns his back
to my residence

Issa

standing as a windbreak
for the suckling
the scarecrow

Issa

the full moon
  stoically
the scarecrow stands

Issa

this old body
even before the scarecrow
I'm embarrassed

Issa

after becoming acquainted
  I then depart
with the scarecrow

Izen

from the moment it's built
  it's old
the scarecrow

Jofū

autumn deep
  a scarecrow wears
  a robe of leaves

Otsuyū

as I walk
a distant scarecrow
walks with me

Sanin

a fallen scarecrow
above his face
    rises heaven

Sanki

the sparrows move
  from scarecrow
to scarecrow

Sazanami

when it rains
the scarecrow looks
just like a man

Seibi

a scarecrow
his bow drawn
toward someone else's field

Setsueyo

resembling men
scarecrows on a moonlit night
evoke pity

Shiki

rice-sparrows
shot by the scarecrow
    fade into the sea

Shiki

to the setting sun
even his face indifferent
the scarecrow

Shirao

come dusk
his face snow-white
the scarecrow

Takaio

picked up
as I walked down the road
I've made a scarecrow

Tōrin

right under his foot
the beans get stolen
the scarecrow

Yayū

spring storm
the scarecrow's hat
   off with the birds

Ernest J. Berry

never an update
from our on-the-scene scarecrow
  filled with last year's news

David S. Bloch

crows leaving
as quickly as they land
—I am the scarecrow

Tom Clausen

daring the crows
overstuffed scarecrow
arms akimbo

Wanda D. Cook

October morning—
making scarecrows
from old clothes

Stephanie Curson

in the neighbor's field
wearing a different coat
the same old scarecrow

L. A. Davidson

after the foliage
the weathered scarecrow
kneeling

Raffael DeGruttola

scarecrow
the first snow
slips from his hat

Peter Duppenthaler

thawing field:
the scarecrow leans into
a puddle of stars

Frederick Gasser

cornfield at midnight:
  this headless scarecrow
    is glazed with moonlight

Lloyd Gold

too wet to sow
the scarecrow
thins in rain

LeRoy Gorman

November morning
scarecrow's faded clothes
the frosted garden

Ruth Holzer

the scarecrow
moving backwards —
autumn rain

Gary Hotham

still smiling
in the November snow—
the orchard scarecrow

Elizabeth Howard

tattered scarecrow
in the center of a storm
rattling cold

Page M. Howell

seeding time
the farmer dresses the same
as the scarecrow

Jim Kacian

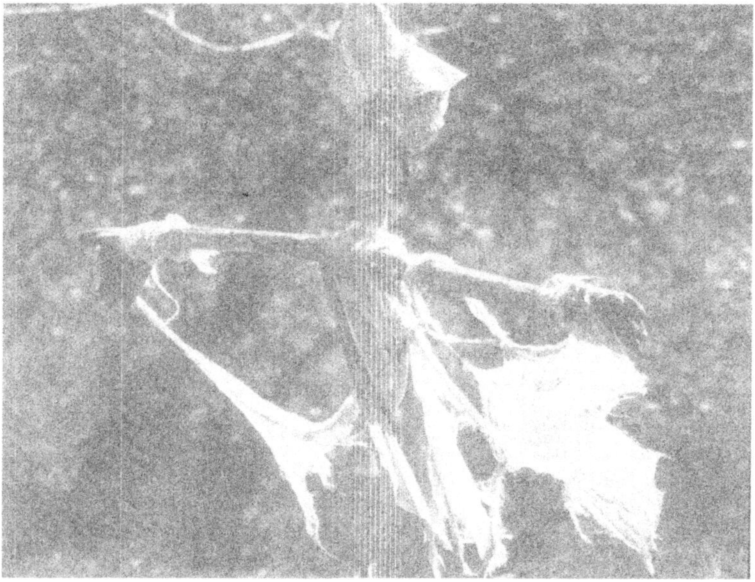

reamins of a scarecrow
remains
of a cornfield

Michael Ketchek

pink Oshkosh overall
on a small scarecrow
autumn rain

Joann Klontz

spare change
in the scarecrow's pocket—
pumpkin seeds

Willard Klontz

one step closer
  a crow caws
    at scarecrow

Alex Kowalczyk

midsummer heat
  scarecrow
with drooping arms

Robert Kusch

murky night —
pointing the way home
a scarecrow

Watha Lambert

Halloween —
even the scarecrow
wears a mask

Dorothy A. McLaughlin

snow-covered field
a single lump marks
the buried scarecrow

Andrea C. Missias

A Scarecrow Sequence

even the scarecrow
would be good company
this morning of fog

The familiar scarecrow, in his old hat and baggy suit, would be a comforting sight compared to the nothingness of fog. Loneliness teaches toleration. To quote Milton: "Who can enjoy alone?"

A
windy
day

speaking
to
the
bobbing
head

of
a
scarecrow

Not wishing to appear less polite than the scarecrow, the haiku poet calls out a friendly greeting.

Brushing the stick arm
of the scarecrow—
all my corn scattered

There is a Turkish proverb: "A weapon is an enemy, even to the owner."

A cold night—
borrowing the scarecrow's coat
to moon-watch

"Borrowed garments never fit well," but it is the nature of human beings to put their own comfort before the needs of others.

With a feather
in his hat—
how youthful the scarecrow

Human scarecrows use all the trappings known to man to appear younger and more attractive, so why shouldn't the inhabitant of the corn field sport a jaunty hat?

Somewhat tipsy tonight
the scarecrow
moon-pointing

Is the scarecrow intoxicated, or has the summer rain loosened the dirt, so that he has tilted sideways? The haiku poet chooses to see only one stick arm—the one pointing upward.

Everything bending
before the wind
except the scarecrow

In contrast to the French proverb: "It is better to bend than to break," the rigid stick-back of the scarecrow will never bend to the will of the wind like the grasses.

A
full
moon
tonight

what
thin
shadows
we
make

the
scarecrow
and
I

What thin shadows we make in this earthly existence. How often they resemble the silhouette of a scarecrow.

Going inside
by the fire —
taking the scarecrow with me

A delightful illusion in which the lone scarecrow is brought in to share the warm fire, or is it because he would be good company as he might havebeen on a morning of fog?

— *Frances Brandon Neighbors*

a herd of hollow tracks
and only one shepherd
the scarecrow

Ana Rosa Nuñez

the hole
where a scarecrow stood —
snails

Fumio Ogoshi

rural snow country
a valley of scarecrows
with day-glo T-shirts

Brent Partridge

worldly scarecrows
sporting cone-shaped hats, straw hats,
Amish bonnets

Francine Porad

my neighbor's scarecrow
lost among the sunflowers
of early August

Linda Porter

wet day
under soggy felt hat
the scarecrow

Patricia Prime

late fall
a scarecrow hunches over
a schoolyard garden

Anthony J. Pupello

a scream
in the scarecrow's mouth
of winter wind

William M. Ramsey

field mice—
the scarecrow's head
drooping lower

Carolyne Rohrig

farm auction—
clothes too threadbare
for the scarecrow

Sharon Lee Shafii

for a bit of dash
  I tied a red necktie
    on the scarecrow

Ossie F. Tranbarger

harvest moon —
a baleful scarecrow
lying in the field

Charles P. Trumbull

scarecrow ready
by this corn-planted patch
not a bird in sight

Anna Vakar

summer scarecrow
with fireflies for eyes
  winking through the night

Alice M. Ward

where the scarecrow
used to stand
        a snowman

Mark Arvid White

looking through the hole
in the scarecrow's hat
the fledgling sparrow

Paul O. Williams

his and hers scarecrows
stand tall—
old couple weeding

Ruth Yarrow

lightning
scarecrow
fireflies

Arizona Zipper

Farewell to the Scarecrow

The scarecrow seems to be one of the "sacred cows" of Japanese poetry. We meet him from the earliest translations into English. Torin, Nyofu, Sanin, Issa, Choi, Basho, Buson — all the great and near-great wrote about him, over and over. He may be found at all times of the day, in all seasons, in every stage of dress and deshabille. He stands as a symbol of the state of mind which loves nature more than man. He is the picture of compose through all vicissitudes. This is the "non-man", the ubiquitous "now", met in a snowfall, at bird-nesting, in the harvest time. He leans, he sags in rain and wind; he does not disdain the casual decorations gestowed by his deriders.

The scarecrow was once an integral part of the Japanese landscape, much as the rice fields which were his sole reason for existence. He still may be found presiding over som epoor man's bit of isolated property — especially in the more remote sections of the islands. But he is rapidly become superannuated and replaced by a variety of more modern devices — such as strings of dancing foils, or mechanically activated gadgets. These can be controlled from the sidelines and only need to be placed during their working hours — the harvest time. Some fields are nowadays protected by sound-making instruments, which work well as protection from the rice-birds, even though they may annoy the neighbors.

In our United States, the scarecrow has largely become an anachronism. The vast majority of our citizens — even in the countryside, have scarcely seen one within recent years. Scarecrows seldom served other purposes than guarding of a corn patch. There the various members of the crow family were assiduous gleaners — both of the newly sown seed and of the ripe and hardened grain. After it was one sprouted, the kernels held less appeal.

In pioneer times when seed was likely to be scarce in spring, and when the labor of hand-planting was a major consideration, the lonely straw man (or woman at times) — left behind as protection — did help in confusing the spring-hungry or late-fall ravages of "Crow".

Now that America has become the land of plenty, the corn crop far outstrips the demand. Its price has to be supported by the government. A few grains or a few hundred thousand stolen are no longer a serious matter. Human labor costs involved are so great, the growing of grain crops has had to become highly mehanized. Hybrid breeding has created giant stock. Machines have leveled hills, erased hedgerows, filled ditches and wiped out boundaries. Aross fields the activity of planting and sowing proceeds in great waves; acres have become green oceans; corn grows in mile squares. Seldom nowadays do we see little fields in late autumn, where drying ears remain in corn stalk tepees. The harvester's swing through stalks higher than the "elephant's eye"; poor scarecrow would indeed be lost.

There is no need∏no place for Mr. Scarecrow. It is a reasonable suggestion that he be retired from American poetry—except as a symbol of olden pioneer days. He certainly does not belong to modern living or modern poetry.

—*Ruth Latta*

Acknowledgments

Illustrations to this book are from *Ephemeral Folk Figures: Scarecrows, Harvest Figures and Snowmen*, by Avon Neal with photographs by Ann Parker (Clarkson N. Potter, Inc., New York 1969).

John Stevenson's foreword "The Scarecrow in our Haiku" was written expressly for this volume.

Yamamoto Kenkichi's "History of *Kagashi*" is taken from *Nihon Dai-saijiki* (Kodansha, 1983). Translation is by Hiroake Sato expressly for this volume.

Translations of Japanese haiku are by Hiroake Sato.

"A Scarecrow Sequence" first appeared in *Haiku West* 8.1 (July 1974).

"A Farewell to the Scarecrow" first appeared in *Haiku West* 3.2 (January 1970).\

Made in United States
North Haven, CT
10 November 2026

22088260R10075